THE GORILLA GUIDE TO...

Modern Storage Strategies for SQL Server

Written by
David Klee
Microsoft MVP and VMware vExpert

and James Green
VMware vExpert

ActualTech Media

The Gorilla Guide to Modern Storage Strategies for SQL Server

Author:	David Klee, Microsoft MVP and VMware vExpert, James Green, VMware vExpert
Editors:	Hillary Kirchener, Dream Write Creative
Book Design:	Braeden Black, Avalon Media Productions Geordie Carswell, ActualTech Media
Layout:	Braeden Black, Avalon Media Productions
Project Manager:	Amy Short, ActualTech Media

Copyright © 2016 by ActualTech Media

All rights reserved. This book or any portion there of may not be reproduced or used in any manner whatsoever without the express written permission of the publisher except for the use of brief quotations in a book review.

Printed in the United States of America

First Printing, 2016

ISBN 978-1-943952-11-3

ActualTech Media
Okatie Village Ste 103-157
Bluffton, SC 29909
www.actualtechmedia.com

About the Authors

David Klee,
Founder and Chief Architect, Heraflux Technologies

David Klee is a Microsoft MVP and VMware vExpert with over seventeen years of IT experience. David spends his days focusing on the convergence of data and infrastructure as the Founder of Heraflux Technologies. His areas of expertise are virtualization and performance, datacenter architecture, and risk mitigation through high availability and disaster recovery. When he is not geeking out on technologies, David is an aspiring amateur photographer. You can read his blog at davidklee.net, and reach him on Twitter at @kleegeek.

David speaks at a number of national and regional technology related events, including the PASS Summit, VMware VMworld, IT/Dev Connections, SQL Saturday events, SQL Cruise, PASS virtual chapter webinars, and many SQL Server and VMware User Groups.

James Green,
Partner, ActualTech Media

James writes, speaks, and consults on Enterprise IT. He has worked in the IT industry as an administrator, architect, and consultant, and has also published numerous articles, whitepapers, and books. James is a 2014 - 2016 vExpert and VCAP-DCD/DCA.

About ActualTech Media

ActualTech Media provides enterprise IT decision makers with the information they need to make informed, strategic decisions as they modernize and optimize their IT operations.

Leading 3rd party IT industry influencers Scott D. Lowe, David M. Davis and special technical partners cover hot topics from the software-defined data center to hyperconvergence and virtualization.

Cutting through the hype, noise and claims around new data center technologies isn't easy, but ActualTech Media helps find the signal in the noise. Analysis, authorship and events produced by ActualTech Media provide an essential piece of the technology evaluation puzzle.

More information available at **www.actualtechmedia.com**

About Tegile Systems

Tegile Systems is pioneering a new generation of flash-driven enterprise storage arrays that balance performance, capacity, features and price for virtualization, file services and database applications. With Tegile's line of all-flash and hybrid storage arrays, the company is redefining the traditional approach to storage by providing a family of arrays that accelerate business critical enterprise applications and allow customers to significantly consolidate mixed workloads in virtualized environments.

Tegile's patented IntelliFlash™ technology accelerates performance and enables inline deduplication and compression of data so each array has a usable capacity far greater than its raw capacity. Tegile's award-winning solutions enable customers to better address the requirements of server virtualization, virtual desktop integration and database integration than any other offerings. Featuring both NAS and SAN connectivity, Tegile arrays are easy-to-use, fully redundant and highly scalable. They come complete with built-in snapshot, remote-replication, near-instant recovery, onsite or offsite failover, and VM-aware features.

For more information, visit *www.tegile.com*

Table of Contents

Chapter 1:
Introduction to SQL Server and Storage **13**

    Data and Infrastructure Intersect 14

        The Storage Layer. 16

        The Physical Server 17

        Virtualization 17

        Interconnects and Networking. 18

        The Operating System and Applications 18

        Database Servers 18

    Modern Data Center Challenges 19

        Complementary Technologies. 20

Chapter 2:
SQL Server I/O **21**

    SQL Server I/O Performance 22

    SQL Server Workload Patterns. 23

    Architecting the SQL Server I/O Stack 25

    SQL Server and Virtualization 27

    SQL Server Instance 29

Chapter 3:
Storage Considerations with SQL Server. **33**

    Performance Metrics. 33

        Latency 34

        IOPs 34

        Throughput 35

        Infrastructure Layer Metrics 35

        Maximums vs. Steady State. 37

    Cache vs. Primary Storage 37

    Disk Configurations. 39

        Traditional. 40

        Tiered 40

        Hybrid 41

All-Flash 42
Efficiencies and Data Reduction 43
Compression 43
Deduplication 44
Data Savings 44
Metadata Management 45

Chapter 4:
Business Continuity & Disaster Recovery 47

BC and Disaster Recovery By the Storage 47
RAID 48
Snapshots 49
Replication 50
BC and Disaster Recovery by SQL Server 51
Database and Transaction Log Backups 51
Log Shipping 51
Mirroring 52
Availability Groups 52
Competing or Complementary? 52

Chapter 5:
SQL Server Considerations 55

SQL Server Licensing Reduction 55
SQL Server vs. Array Features 58
Compression 58
Transparent Data Encryption 59
Partitioning and File Groups 60
Buffer Pool Extensions 61
Monitoring, Management, and Support 63
Monitoring 63
Management 63
Support 64

Chapter 6:
Storage Best Practices **65**
    Hardware and Bottleneck Detection. 66
        Storage Bottlenecks 66
        Operating System Bottlenecks. 68
    Block Size. 69
    Tuning for Flash 71
    SQL Server Storage Requirements and Budget. 73

Chapter 7:
Modernizing SQL Server **77**
    Features in the Latest Version 77
        In-Memory OLTP. 77
        Stretch Database 78
        Azure Integration 78
        Buffer Pool Extensions 79
        Clustered Columnstore Indexes. 79
        SMB3-Based Network Shares 79
    Benefit Analysis of Upgrading 80
        Support 80
        New Features 81
        Widespread Business Benefits. 81
    Maximizing SQL Server with Flash 82
        Perception and Justification 82
        Adoption. 83

Gorilla Guide Features

Food For Thought
In the these sections, readers are served tasty morsels of important information to help you expand your thinking.

School House
This is a special place where readers can learn a bit more about ancillary topics presented in the book.

Bright Idea
When we have a great thought, we express them through a series of grunts in the Bright Idea section.

Dive Deep
Takes readers into the deep, dark depths of a particular topic.

Executive Corner
Discusses items of strategic interest to business leaders.

Introduction to SQL Server and Storage

The volume of data retained and accessed by today's organizations is exploding at an unprecedented rate. Business expects the IT department to keep data online and accessible indefinitely.

Smartphones and unlimited apps require data at speeds and volumes that have the carriers bursting at the seams. The unstructured data recorded by the Internet-of-Things phenomenon is in its infancy but already provides incredible data to the manufacturers. "Big Data" and business intelligence require the storage and immediate retrieval of terabytes, or even petabytes (and beyond), of data.

So of course businesses struggle to keep up. All of this data has to be stored somewhere, somehow. The space must be available to store the data on. The speed must be there to return the data to the users at the speed of business, which is always "right now," with no delay. This data-storage platform must also be readily available in spite of hardware failure, natural disasters, or human error.

The systems intended to deal with this data have been historically very complex. Full-time infrastructure and database administrators spend careers protecting the data and the infrastructure around it. Each system component requires planning and testing to make sure that a failure will not hurt the business. Plus, each layer requires in-depth architecture and careful planning to ensure that the stack will perform at its fullest potential. This means best-of-breed components, cutting-edge techniques, and highly trained staff are needed to engineer these platforms.

As a result, the investment by the business is quite significant. After all, the data it protects, and the systems that it resides on, make up the core of most businesses. If the data is unavailable, or even worse — it's lost — the business is doomed. Sometimes businesses will actually fail because of the loss of data.

Therefore, it is the administrator's responsibility, first and foremost, to ensure that data is not lost, no matter what event occurs. Ensuring that the data is presented to the application in a timely manner is a very close second.

Data and Infrastructure Intersect

As shown in **Figure 1-1**, many layers exist in the core of these data platforms. One traditional challenge with these complex infrastructures is how the layers interact with each other, and how they are managed as a whole. A best practice for one layer might be a horrible idea for another layer. An application-specific practice for a layer might hurt the effectiveness of others.

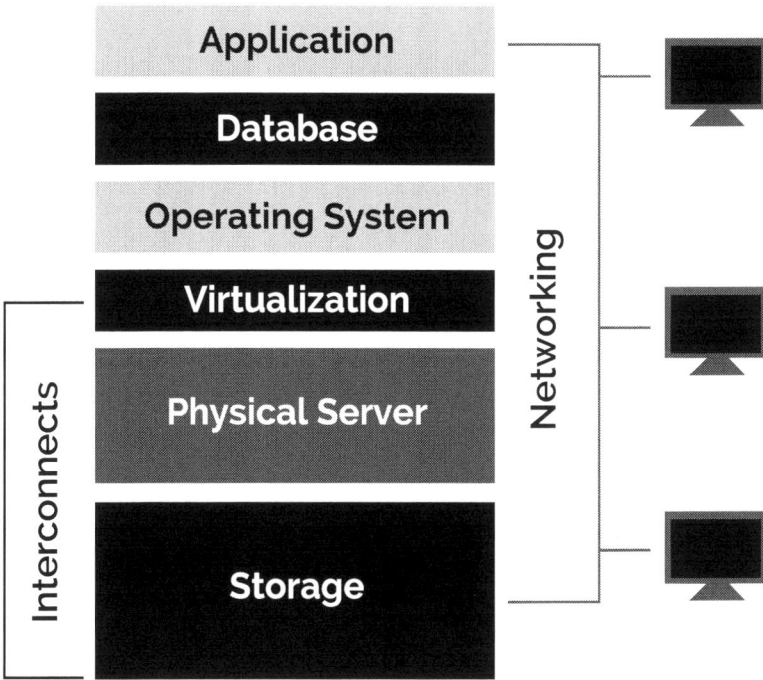

Figure 1-1: The layers of a data center

Learning how these layers interact, complement each other, and potentially compete with each other is vital to architecting the modern on-premises data center.

Adding further complication into the mix is the continued growth of the public cloud. Permutations of public cloud offerings can complement and even replace some of the layers in this infrastructure. Applications and/or databases-as-a-service can remove the requirement to manage the operating system underneath them. Applications, databases, and VMs can be replicated to the public cloud for DR purposes, or can even replace the need to have these components on-premises altogether.

The Storage Layer

The bottom layer in this data center model (**Figure 1-2**) is the storage subsystem. This device is arguably the most critical layer in the data center. It physically stores the data into bits and bytes onto persistent storage, which is usually made up of spindle-based magnetic or solid-state flash hard drives.

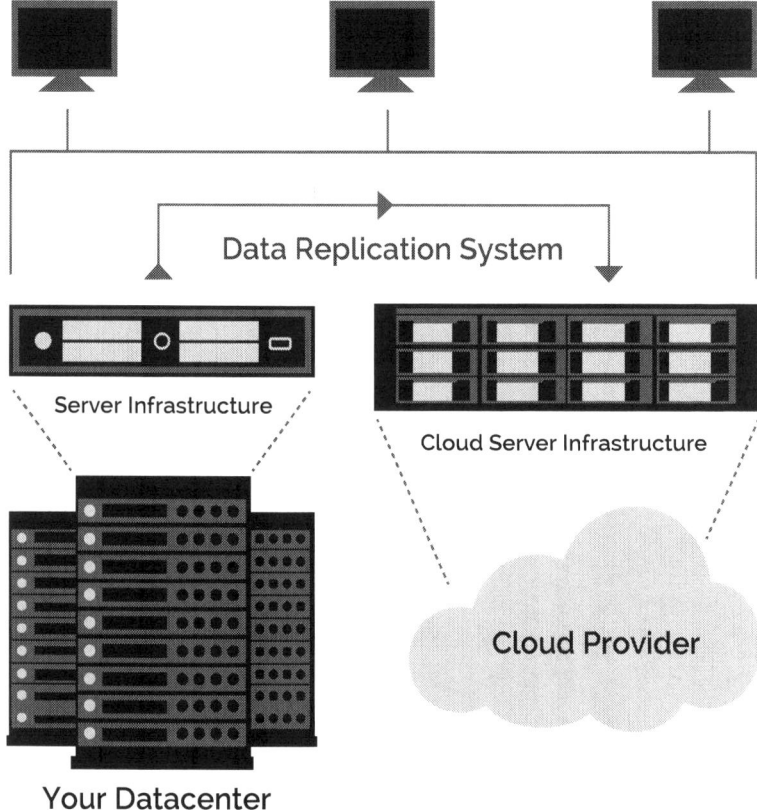

Figure 1-2: Hybrid Cloud Datacenter

While the storage can be contained within a physical server, this configuration presents a single point of failure in that physical server. It is usually configured as a shared device in a Storage Area

Network, or SAN. Usually, SANs are in place for simplifying management and reducing the footprint in the data center, and not necessarily because of their speed. SANs are designed to be robust and protect the data at all costs, and to protect against failures.

Ordinarily, the SAN is the slowest layer of the infrastructure stack. SANs usually measure response time in milliseconds, while server memory or CPUs are measured in nanoseconds.

The Physical Server

The next layer is the physical server, which contains the rest of the compute hardware (CPU, memory, and interconnect adapters). This layer is where the actual work of the application is performed. The data is still stored on the SAN.

Virtualization

The physical server normally contains some type of virtualization, which is an added layer treated as an extension of the physical server.

This layer, called a *hypervisor*, allows multiple compartmentalized operating systems and their respective applications to coexist independently on the same hardware. The virtual machines submit requests for compute resources, and the hypervisor coordinates the access to the physical compute resource to fulfill the request.

By allowing more than one of these virtual machines to run on the same physical server at the same time, the efficiency of the environment improves, and things such as the actual physical server count, rack space, power, cooling, and interconnect cabling, all get reduced. This equipment reduction saves overall data center costs.

While not mandatory, most data centers use virtualization to encapsulate workloads and improve the agility of the data center in order to increase the responsiveness to an ever-changing world.

Interconnects and Networking

The physical server must be connected to the SAN and to other servers through an interconnect, usually fiber optics or high-speed networking, which provides one or more layers of interconnect switching.

Data centers commonly have a dedicated equipment to handle storage communication, and another set to handle application communication so that one does not hinder the performance of the other.

The Operating System and Applications

The logical server, be it physical or virtual, contains an operating system where different applications servers are to be installed onto. Some of these applications servers contain applications which present the data to the end users, such as a web server, while other applications servers are the database servers that contain the data, such as a SQL Server.

Database Servers

Data must be stored in a system that is designed to store and retrieve the data efficiently. The database server is the gateway to the critical business data that is stored on the SAN. Microsoft's flagship data platform, SQL Server, is one of the most widely used database engines in the world, and is at the core of this guide.

Modern Data Center Challenges

Data center topologies are amazingly complex. No two data centers are the same, but all face common challenges with storing and retrieving large volumes of data. These challenges include:

- **Data age.** Businesses demand that a lifetime of data remains accessible. This means that the data must all be online, all the time.

- **Cost.** Data volumes are growing exponentially but contrary to popular myth, storage prices are not getting any cheaper. SANs are sometimes prohibitively expensive, and are almost always the most expensive component of a data center.

- **Complexity.** The complexity of managing data increases dramatically as the volume and performance demands increase. The agility and flexibility in the environment is reduced, and grows in management costs at an unsustainable rate.

- **Scalability.** Infrastructure bottlenecks can exist in any data center. As the volume of data goes up, certain components just cannot handle the load and present pain points in different ways. Sometimes the applications slow down, the storage fills up, or the infrastructure becomes unreliable. The business suffers from the lack of predictability in the environment.

- **Data sprawl.** Multiple copies of data can litter the environment. For example, development and test copies of data increases the volume of data an organization is forced to maintain, and it drives up the cost of data storage even more.

- **Interaction.** Processes and management techniques for one layer can conflict with the best practices for another layer. The impact of these competing processes can be performance penalties, more complex configurations, or worse, even system outages.

Complementary Technologies

A working knowledge of the basics of each layer is critical to properly understand how to leverage each layer and improve the reliability and performance of the others.

Two of the most challenging layers - storage and the database - lie at the heart of the application stack. Fortunately, these layers do not have to be at odds with each other. Properly architected, these two layers can be quite complementary and present a technical solution that improves the business's chances at succeeding in today's competitive world. Each layer and how they interact will be discussed in depth in subsequent chapters.

Up Next

With an understanding of the ways the data and storage layers interact in the modern data center, let's move on to discuss some of the intricacies of the database engine and the ways it uses the storage layer to function.

2

SQL Server I/O

Microsoft's flagship database engine, SQL Server, is arguably one of the best enterprise relational database engines out there. SQL Server is underneath many of the critical applications at businesses around the world.

At the heart of SQL Server are two layers: the query and storage engine. The query engine interprets an application's request for the retrieval of specific data. The storage engine then takes this specific request from the query engine and fetches the appropriate data from disk.

The storage engine uses a portion of Windows Server operating system memory as an I/O read cache to store the working set of data in while it finishes its retrieval process. It also uses this memory as a buffer to improve the performance of queries that fetch frequently read data.

SQL Server I/O Performance

A normal SQL Server can be one of the largest I/O consumers in your server environment. A SQL Server with a poorly designed database, bad queries, or improperly managed maintenance strategy can go well beyond "busy." It can overwhelm your storage, which can cause the speed of the I/O operation, or latency, to skyrocket, slowing down the operation.

Why?

Because when a database receives a request for a certain set of data, the query engine determines what data to fetch and instructs the storage engine to retrieve it. The storage engine then goes and carries out the request. If the query asks for more data than it needs, such as unnecessary table columns, the storage engine has to retrieve more blocks of data from the storage. Sometimes the request could require certain filtered data, such as the data between two dates, or the data that pertains to a specific purchase order number or customer. That means the query engine might have to scan all the data to find the specific records if the right indexes are missing. This is not ideal. Other times, different tables might be joined to fulfill the request, such as a request to get all orders that a particular customer has placed in the previous year.

Any of these operations can cause elevated I/O consumption. Whenever these operations are processed, a single query can cause gigabytes (or more) of data to be read from disk but end up only needing to return one or two records. The speed of the SAN is critical to the overall runtime of the query.

SQL Server Workload Patterns

Each of these requests from the SQL Server storage engine can generate a different workload pattern on the storage layer. These patterns impact your storage performance if the infrastructure configuration is sub-optimal for the workload type. Fortunately, there are similarities between workload types and patterns will then emerge from the workload. The storage array can then be tailored to best fit the workload properties.

Figure 2-1 demonstrates just some of the various workload patterns that SQL Server will use as it accesses the storage layer throughout normal daily operations.

Workload Type	Block Size	Disk Pattern
OLTP	8KB / 64KB	Random read/write
OLTP Read Ahead	8KB - 1MB	Sequential read
OLTP Table Scans	512KB	Sequential read
Transaction log commit	60KB	Sequential write
Transaction log read	120KB	Sequential read
Database backup	1MB	Sequential read
Bulk load	256KB	Sequential write
SSAS Cube Workload	32KB	Random read/write

Figure 2-1: Workload Characteristics

Keep in mind that the intensity of these workloads is largely based on the application demands of the database and it can be quite severe at times. Some of these demands include:

- **Routine maintenance operations.** Routine maintenance operations can also create significant load against the storage layer. SQL Server databases need periodic index and statistics maintenance to stay optimal. This index maintenance can create an intense window of random read and write activity.

- **Routine database integrity checks.** Routine database integrity checks also create a large amount of read activity.

- **Other tasks.** Activities such as nightly data loads and analytics processing, can generate quite a bit of activity too.

These workloads and tasks, that come from seemingly every direction, can bury even the highest-performing storage if not properly managed.

Work with your database administrators (DBAs) to map out each one of these routine activity windows per server, and map it across the entire environment. You can see an example of these activities in **Figure 2-2**.

Task	Start	Avg. End	Frequency
VM-level backup	2 AM	2:30 AM	Nightly
DB full backups	3 AM	4:10 AM	Nightly
DB transaction log backups	12 AM	(20 sec)	Every 5 Mins
DB index maintenance	10 PM	11:30 PM	Nightly
DB integrity checks	6 AM	9:20 AM	Every Sunday
Data bulk load	6:45 PM	7:30 PM	Nightly
Inventory refresh	7:45 PM	8:20 PM	Nightly

Figure 2-2: Routine Scheduled Tasks

The activity patterns on the SAN can be directly traced back to these windows of activity, and the results might surprise you. DBAs might not realize that workload runtimes might be overlapping, or that the concurrent tasks might compete for I/O and slow down the overall performance of the environment.

Architecting the SQL Server I/O Stack

Let's dive deeper into the infrastructure between the SQL Server query and the depths that it must go to get the bytes of data from disk.

As shown in **Figure 2-3**, the bottom layer is the storage layer. This device consists of spindle-based magnetic disks, solid-state flash memory disks, or a combination of the two. These disks are logically grouped into one or more pools of disk, configured in a fault-tolerant manner so it can withstand drive failure without losing data.

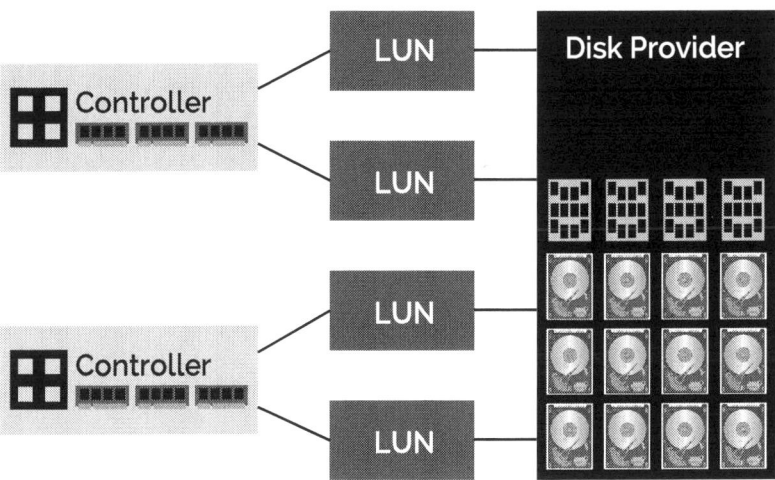

Figure 2-3: SQL Server I/O Stack

From there, one or more SAN controllers act as an intermediary between the outside world and the storage disks. Their job is to accept an I/O request, send the request to fetch the block of data from the disk pool, and return this data out the way it came. These controllers contain network or fiber optic adapters that connect the controller to the outside world, and contain their own CPUs and cache memory, which act as an I/O read-and-write buffer to boost the overall performance of the array.

The SAN controllers connect into the storage interconnects switches. This layer allows the storage controllers to be connected to many servers or other devices on the network. Common interconnect architectures can leverage fiber optics or traditional networking layers. Furthermore, a controller can have more than one interconnect port on the device. The additional ports can help to load balance across the available end-to-end connection paths, and can also reduce downtime if one of these ports were to fail.

These ports are also rated for a maximum data transmission speed. These speeds are important because if the SAN is faster than the interconnect's end-to-end overall path, the performance of the storage — as the endpoint experiences it — might not reach its full potential and could cause slower-than-expected performance.

At the other end of the interconnects are the physical compute servers. The interconnects connect into these physical servers through devices such as host bus adapters (HBAs) in the case of fiber, or network interface cards (NICs) for Ethernet connectivity. As with the SAN, more than one can be used to provide greater performance and resiliency against equipment failure.

These compute nodes also contain the processing portion of the infrastructure stack: CPUs and memory. The CPUs are configured with multiple CPU-processing cores on a single chip, and multiple sockets for these chips can be used in tandem in the server.

Memory is then placed next to each of the CPU sockets to improve the performance. This configuration creates non-uniform memory access (NUMA) nodes, or a grouping of CPU sockets and a bank of memory.

SQL Server and Virtualization

Virtualization is now everywhere in data centers worldwide, and chances are, your SQL Servers are probably virtualized by now, too. That's a good thing. The added benefits to the data center are tremendous, and the agility that virtualization brings to your enterprise is game-changing.

However, virtualizing SQL Server requires quite a bit of tender loving care to ensure that the performance of the database stays high.

The primary challenge of virtualized SQL Server comes from how virtualization creates a shared-everything platform of compute resources. Applications like SQL Server do not like to share. They are extremely sensitive to any delay of the infrastructure underneath.

Plus, infrastructure admins usually do not have much hands-on experience with SQL Server. Practices that work great for data center consolidation can greatly penalize the business-critical application performance.

This presents a problem of epic proportions.

Care must be taken to minimize the impact of multiple virtual machines (VMs) trying to access the same physical compute resources through the hypervisor. If one VM consumes all of the resources on a host, the impact on the other VMs on the same host can be quite negative.

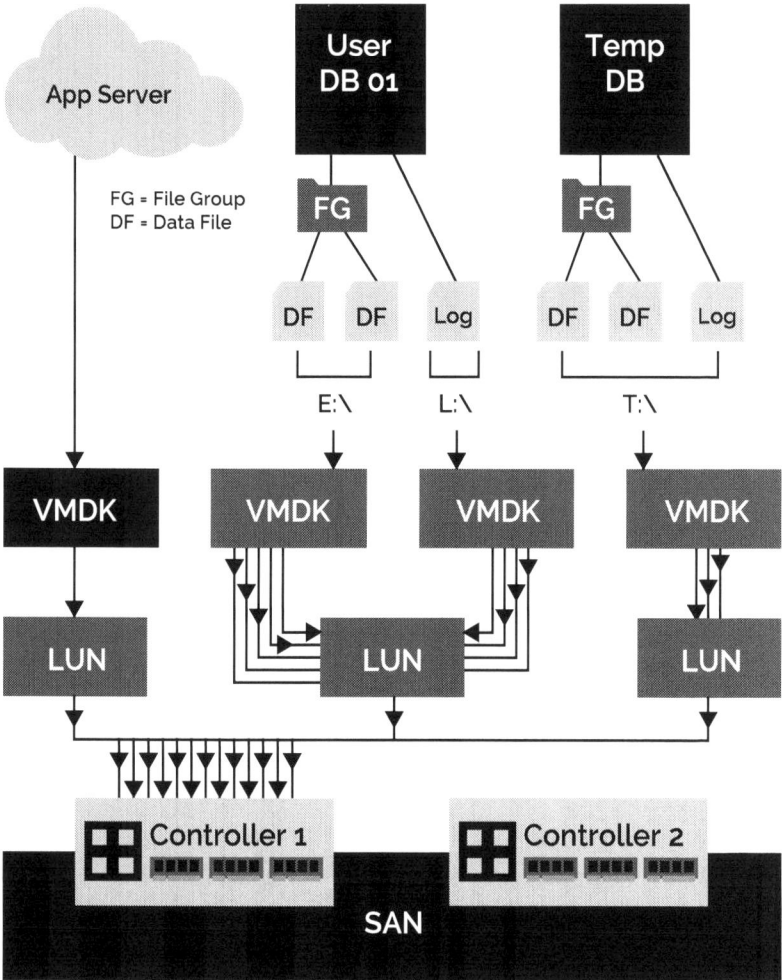

Figure 2-4: Heavy I/O Traffic Causes Contention

For example, if one SQL Server is running a database integrity check, which is very I/O intensive, the SAN LUN and controller that the data is traversing could become quite active and busy.

The added load could cause the physical HBA or network adapter to reach its maximum throughput. In some cases, no room exists on this interconnect for the other VMs to perform their normal duties. As a result, the other VMs I/O request is backed up in the queues inside the hypervisor, and the application experiences the impact of this in the form of high disk latency.

SQL Server Instance

The SQL Server installation, or instance, is a logical hierarchical structure, as shown in **Figure 2-5**.

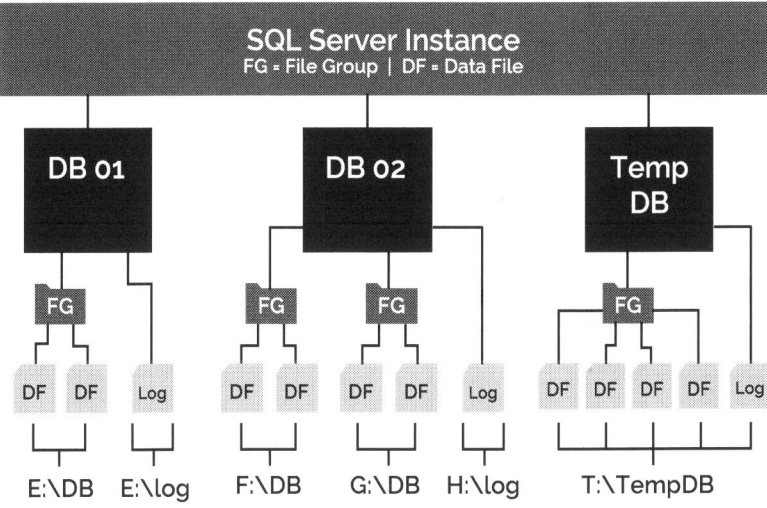

Figure 2-5: SQL Server Database Structure

A SQL Server instance can contain one or more databases. Each database contains one or more logical file groups, which themselves contain one or more data files which store the information in tables. These files can be placed on one or more logical drives or mount points inside Windows Server.

Slow round-trip latencies in the storage layer's response to SQL Server's I/O requests can dramatically slow down the performance of the database queries.

In addition to the database data files, a transaction log file is used to store the record of times that are added into the database, or store the details of data that change or are removed.

Any change into the database is written to this log file, and one the database command will not finish until this transaction has been written to disk. The database depends on this log to provide a record of what has changed and when, and it must be in perfect synchronization with the data files for the database to operate normally.

If the storage layer cannot write this stream of data in the log file fast enough, the performance of database change operations is sure to suffer.

A database specially designed for use by the instance for temporary objects, called *tempdb*, is also present on each instance. This database is used for tasks such as index maintenance, database integrity checks, and query sort operations. This database's performance characteristics vary wildly by application and instance, and special care must be taken to ensure that this database's performance does not negatively impact the performance of the task that depends on it.

Up Next

With SQL Server's I/O architecture detailed, we can now move on to discussing how the two layers — SQL Server and storage — complement (and potentially contradict) each other in the enterprise.

3

Storage Considerations with SQL Server

Knowing how dependent SQL Server is on the performance of the storage, certain architectural considerations must be reviewed and analyzed (especially by any organization with enterprise data needs) to determine the impact the storage performance and the SQL Server demands will have on the performance of the application.

Performance Metrics

First, let's define how to objectively demonstrate the storage performance. These metrics will help you understand the steady-state performance of the storage, and will help you track and trend storage performance to see it's running as expected.

The metrics that you should care about are:

- Latency
- I/Os per second (IOPs)
- Throughput

 **Workload Separation**

For all three of these metrics, you should separate read and write operations because the performance characteristics of these two operations can be very different.

Latency

The amount of time taken for the round trip of one storage request from the operating system layer to the storage and back is called *latency*.

On most SANs it is measured in milliseconds. Newer all-flash SANs can have performance so high that the performance can be less than one millisecond, even into the microsecond range.

For storage performance, lower latency is better, and the lower the latency, the faster the SQL Server can complete the I/O request.

IOPs

I/Os per second, or IOPs, is the number of outstanding requests from the servers that the storage can handle at any given time.

The combination of interconnect speed, storage controller performance, cache memory, and disk configuration help to dictate the maximum number of IOPs that the SAN can deliver.

Higher available peak IOPs is also critical to maintaining the SQL Server's peak performance. For highly concurrent and active SQL Servers, IOPs values and the latency per I/O can determine the scalability and high-end performance of the entire application stack.

Keep in mind that even if the storage can deliver an exceptionally high number of IOPs, the application might not demand this from the storage layer. Some SQL Server workloads do not have a workload high enough to push the IOPs demand very high.

However, even if the demand in IOPs is low, low storage latency will help improve the performance of the database requests.

Throughput

Throughput is simply the server's formatted disk block size multiplied by the IOPs, usually measured in megabytes per second (MB/s). It is regarded as a trusted measurement of performance. However, while throughput is important, latency and IOPs are much more important to the overall system performance than MB/s.

Infrastructure Layer Metrics

When reviewing the performance metrics, make sure that you sample and review the performance metrics from each layer in the system stack.

The performance metrics from each of the following items should be overlaid to determine if the end-to-end performance is operating as expected.

- SAN controller host bus adapters/network interface cards (HBAs/NICs) throughput, by port
- SAN controller CPU performance
- SAN controller cache read and write performance and ratios
- Hypervisor (if applicable) HBA/NIC throughput, by port
- Hypervisor (if applicable) SAN LUN or datastore throughput, latency and IOPs

- Virtual machine (VM), if applicable, disk latency, IOPs, and throughput, split by virtual disk
- Windows Server logical disk latency, IOPs, and throughput metrics, by drive or mount point
- SQL Server disk stall metrics, by database file

Interconnect Bottlenecks

A key point with throughput is that the interconnect from the storage to the server is rated in *maximum theoretical throughput*, or how fast the end-to-end connection can move data. Even if everything is absolutely perfect in the environment, and no overhead is present in the environment, you will not be able to exceed this value of performance on a single interconnect path.

For example, if the storage is connected via a single 8 Gb/s fiber optics (common in data centers), every 8 bits of data gets encoded to a 10-bit payload. The other 2 bits are for data integrity.

Eight gigabits per second, converted to megabytes per second, translates to a theoretical top throughput of 850 MB/s down that individual path.

No interconnect topology is 100% efficient, and overhead comes in the form of switch hops, broadcast traffic, or multipathing overhead, so the expected maximum actual throughput of one-way traffic on this path would be in the neighborhood of 780MB/s.

This speed might be artificially limiting the peak performance of the storage. Additional interconnects with multipathing can (and should) be configured to increase the total overall throughput.

Maximums vs. Steady State

An important distinction must be drawn between the maximum performance an individual infrastructure component can deliver, versus the ongoing "normal" operational range of these components.

Think of it like this. The next time you're in your car, take a look at how high the speedometer goes. When was the last time you maxed out the speedometer? Can your car actually reach the maximum speed listed on the speedometer? Storage performance is the same way. Under normal circumstances, the storage IOPs and throughput demands are usually well under the maximum capabilities of the storage. However, occasionally, some SQL Servers do demand that peak performance.

> **Storage Maximums Testing**
>
> Microsoft's *diskspd* utility can be used to stress test the storage underneath your SQL Server. You can configure numerous parameters to simulate any type and intensity of workload that you need.

Cache vs. Primary Storage

The SAN contains both cache and primary storage drives as part of the basic configuration. Cache is considered transient buffering for both frequently accessed storage blocks and for data write operations. Cache is usually many times faster than the permanent storage layer. The goal is to improve the performance of the SAN while reducing the latencies in reading data from, or writing data to, the persistent storage medium. (**Figure 3-1**)

Figure 3-1: SAN Controller receives a write

Legacy SAN cache uses small amounts of battery-backed memory to buffer these requests, while modern SAN technologies use a combination of large amounts of memory and SSD to increase the performance and capacity of the cache, which in turn improves the overall performance of the SAN.

When data is written to the SAN by a server, the controller accepts the request and writes this data into the controller cache. The SAN then returns an acknowledgement that the write completed successfully. (**Figure 3-2**)

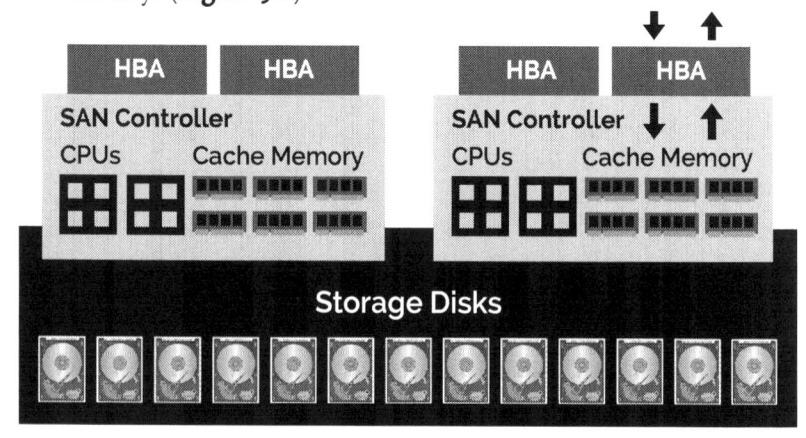

Figure 3-2: SAN Controller acknowledges write

Chapter 3

At this point, the SAN then works to copy the newly written data to the internal permanent primary storage layer. The block of data will eventually be written to the primary storage layer. (**Figure 3-3**)

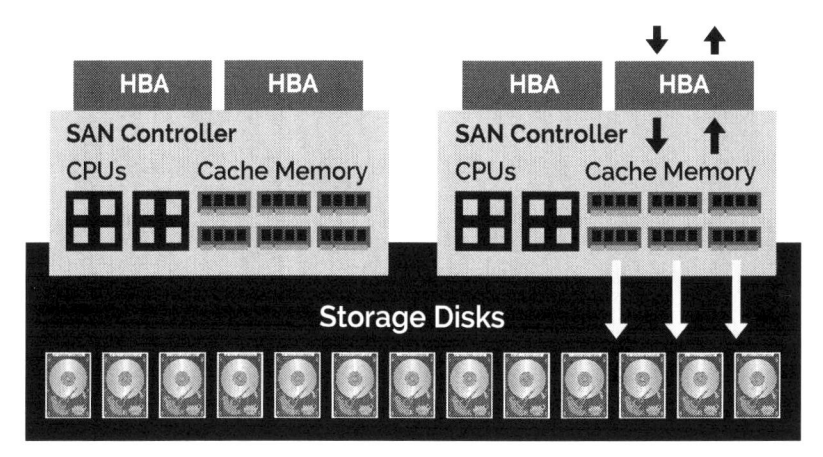

Figure 3-3: SAN Controller flushes write to disk

If the SAN has a large amount of cache, it can retain a copy of it in this layer in case this data is requested in a read operation in the near future.

Large amounts of read cache can improve the overall performance of the SAN if the workload is largely made up of read operations. Most database operations are over 80% read in nature, and large amounts of SAN cache usually improves the overall performance of the database.

Disk Configurations

Modern SANs can come in shapes and sizes big and small, and the performance capabilities range from slow to lightning fast.

Traditional

Traditional SANs contain a number of identical spindle-based primary storage disks and a small amount of cache memory. (**Figure 3-4**) These arrays were historically built for redundancy and resiliency. Improved speed could be introduced with a greater number of spindles and larger amounts of cache memory.

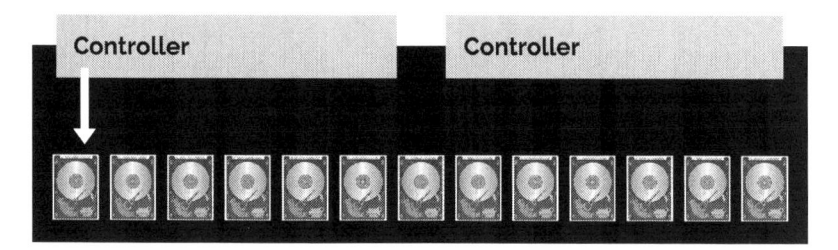

Figure 3-4: Traditional storage array

Tiered

Tiered SANs use multiple groups of disks, each with different performance characteristics. Data is usually placed on the fastest tier of disk first. If the data is untouched for some length of time, the system can relocate the block to a slower (and less expensive) tier of storage during a periodically scheduled maintenance process.

Over time, commonly accessed data is kept at the fastest layer, and archival data sinks to the bottom of the performance. While cost effective, the data movement process can be taxing on the system. Accessing infrequently used data can cause the operation to run very slowly because of the speed of disk where that data is stored. (**Figure 3-5**)

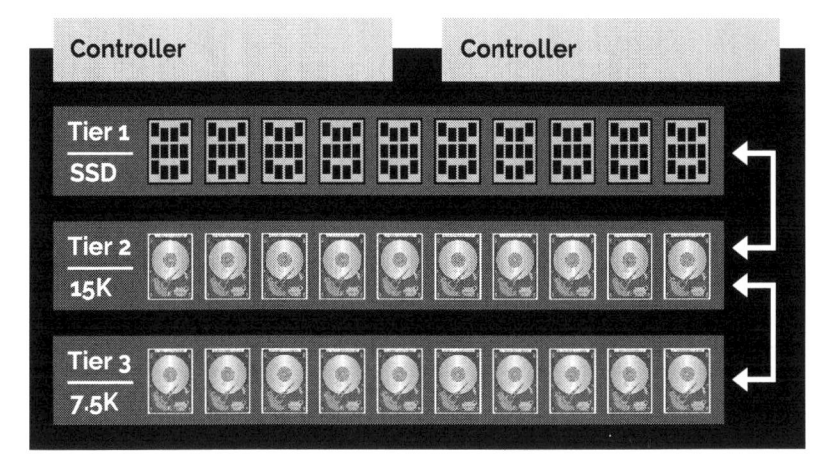

Figure 3-5: Tiered storage array

Hybrid

The tiered SAN model, with the manual movement of data tied to a process that only occurs infrequently during a maintenance operation, has given rise to hybrid SANs.

Hybrid SANs perform a similar type of hot and cold data relocation process to tiered SANs, but do so automatically and around the clock instead of during a particular maintenance window. This improves the overall performance of the SAN.

Hybrid SANs usually also contain large amounts of cache memory to improve the performance of read operations. If an active block of data is constantly re-read from the SAN, the SAN will keep this block in RAM to deliver the fastest possible performance for this operation, as RAM is many times faster than SSDs or spindle disks at this time.

Hybrid SANs also ensure that all write operations are performed on the SSD layer, so that the operation is performed as fast as possible while maintaining the critical integrity of the data being written to disk.

Storage Considerations with SQL Server

Common configurations can include a two-layer solution with a tier of solid-state drive (SSD) disks for very fast write operations and a tier of SAS spindle-based disks for longer-term data placement. (**Figure 3-6**)

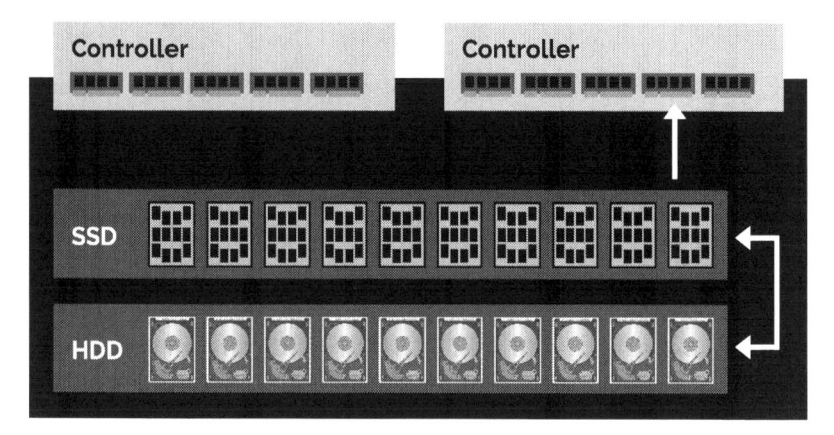

Figure 3-6: Hybrid Storage Array

All-Flash

For the most demanding of workloads, all-SSD (all-flash) storage can be leveraged to provide the fastest possible performance of the shared storage models. While more expensive than traditional or hybrid SANs, all-flash SANs provide the best possible performance. (**Figure 3-7**)

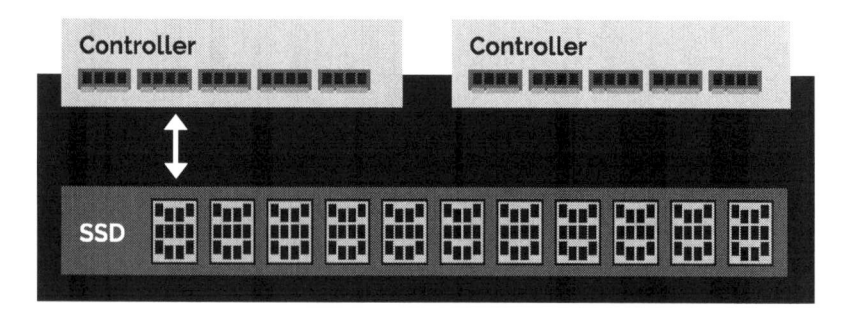

Figure 3-7: All flash storage array

Even with the impressive speed of the all-flash SSD layer, large amounts of cache memory are leveraged in each controller to improve the overall performance of the array that much more.

Efficiencies and Data Reduction

Newer SAN architectures can incorporate features that allow greater efficiencies in the data storage so that more data can be squeezed onto the SAN. Compression can be used to reduce the footprint of the data stored on disk, and deduplication can be leveraged to make the disk storage and retrieval processes even more efficient. The catalog of where the data is stored on where the array is also able to be more efficiently managed, improving the performance as well.

Compression

A block of data might not contain all random and unique data. Usually, empty space and repeating patterns exist in portions of the data. Compression algorithms can be used to compress the block of data as it is passed into the SAN, and decompressed on its way back out. All of these operations occur transparently to the servers reading and writing to the SAN.

This compression algorithm must be efficient, and the storage controller must have enough processing power to handle these requests fast enough, or the overall read and write process can slow down the performance of the SAN. The result is lower latency, IOPs, and throughput performance metrics from the SAN, which, in turn, can slow down the performance of the business applications.

Deduplication

The blocks of data written to disk can also be redundant in nature. Databases from a production server can be backed up and restored onto a pre-production server for testing. These blocks of data are identical in nature, and just take up extra space on disk. As a result, some SANs have deduplication algorithms built into the units to reduce the data written to disk.

If a block of data is to be written, and it is identical to another block, a pointer is written to the original block instead of writing the identical block a second time. The net result is additional space savings on disk.

As with compression, the SAN controllers must be fast enough to perform this deduplication without imposing a performance penalty on the I/O operation.

Data Savings

Either compression or deduplication can reduce the storage consumed on disk, and when combined, can contribute to impressive savings on disk. The net result is that the amount of data written to the storage is smaller than the amount that the server "sees."

Most blocks of data are compressible, and many environments have large quantities of duplicate data. The space savings can be upward of 40% to 70% on disk. The performance of the SAN can improve as well, as the SAN must retrieve fewer blocks of disk from primary storage to fulfill a request.

For example, 20 TB of enterprise data could be placed on a SAN with only 12 TB of actual usable storage, and the compression and deduplication could provide additional storage space to be used by the business.

> **Layers and Roles**
>
> Data reduction is usually more efficiently handled by the SAN than within the server. Communicate with your SQL Server database administrators (DBAs) about the use of compression and deduplication in your environment.
>
> Some DBAs have data compression within the database at their disposal. If they have data compression enabled inside the database, it could greatly reduce the effectiveness of data reduction algorithms at the SAN and lower the performance of the SAN.

Metadata Management

On some SANs, the catalog of where on disk each block of data resides, called metadata, is comingled on the data disks. Every time a block of data is fetched from the SAN, this metadata must be referenced to determine where to fetch that block of data from, which introduces additional overhead on the array.

On other SANs, however, this metadata is handled and stored separately from the data. Tegile, for example, stores metadata on SSD disks and mirrors it in memory, which ensures that the lookup of the data blocks is performed as fast as possible. Any changes to the metadata are written to the SSD metadata disks. The operational speed of managing the metadata separately from the data layer improves the efficiency of the array, which improves the overall performance.

Up Next

Storage architecture matters with SQL Server, and certain SAN features can improve the performance of the database while reducing the space footprint at the same time.

Let's move on to learning different business continuity strategies for both SQL Server and storage.

Business Continuity & Disaster Recovery

The data on your database must be protected from data loss at all costs. Data loss is one of the worst situations a business can experience, and big enough data loss or downtime could actually destroy the business.

Both the SAN array and the database engine provide excellent methods for protecting your data and replicating it to a different array. Sometimes these features are quite complementary to each other, but other times one feature can directly compete with another. Which is the right strategy for your organization?

BC and Disaster Recovery By the Storage

The SAN provides a number of different methods to protect the data against human error and technology failures.

Definitions

First and foremost, let's define some terms about key aspects of business continuity (BC). These items must be identified by the business before any technical solution can be implemented.

Recovery Point Objective (RPO)

Recovery Point Objective (RPO) is the block of data, represented by time, that the business will tolerate losing. For example, if a business sets four hours as the RPO, then — at a minimum — the data must be backed up and successfully replicated every four hours. Reducing the RPO means that data must be successfully backed up more frequently.

RPO strategies become more complex as the RPO decreases.

Recovery Time Objective (RTO)

If you are willing to lose a certain amount of data, how long is the business willing to wait until the rest of the data is restored and accessible? The *Recovery Time Objective* (RTO) is the maximum length of time that the systems can stay offline for a given situation.

Just like with RPO, the strategies get more complex as the RTO gets closer to zero.

RAID

The SAN itself provides some internal protections against data loss. RAID is a key component of all SAN disk configurations. It allows for the data to span more than one internal disk. That way, if a drive were to fail, no data is lost, and the array can (eventually) rebuild the data that was on the failed drive.

RAID types like RAID 10 and 5 have been used for years by storage admins to protect the data, but nowadays the SAN usually has a preferred RAID configuration for its architecture, and you might not want to change it.

Snapshots

SANs also have features that help provide recovery points that can insulate the business against software and user error. LUN-level snapshots are a point-in-time recovery point on a LUN, and can be taken without downtime to the application then reverted back as needed in the event of a problem.

These snapshots can be scheduled to run at routine intervals across the business day, and will clean up after themselves after a fixed window of time has passed.

A real-world example of when snapshots can help you is when a snapshot is taken just before an application upgrade begins. Let's say that during the upgrade, the upgrade process fails, but only after an entire database table has been updated. The only option is to undo everything that changed all the way back to the point just before the upgrade. Without a LUN snapshot, the application owner would have to roll back every change that the failed upgrade process did not clean up, and the DBA would have to restore a database back to the point before the upgrade. Either of these steps could take hours or longer.

The LUN where all of this data resides could be rolled back to that point in time in just seconds, undoing any change with an automated process that is sure to save the business hours. The environment is then reset so that the business can resume operations while they reset and figure out why the upgrade failed.

Types of Replication

Two types of data replication architectures exist for moving data between SANs: synchronous and asynchronous.

Synchronous replication means that for every block of data written to the SAN, the block gets mirrored to another SAN and successfully written to disk before the I/O operation completes successfully. This type of operation is usually used within a single data center because the slower speeds between data centers can reduce the performance of an I/O operation's round-trip than working in the primary data center.

Asynchronous replication means that any data change to the primary array gets queued up and sent over to the other array as fast as possible.

Replication

SAN-level replication is when a SAN LUN is configured to replicate its contents to another LUN on another SAN (either in the primary data center or to a remote device). This process also occurs at the LUN. It can be configured to run synchronously or asynchronously.

Asynchronous replication can be configured for either a constant stream of changed blocks, or to run periodically on a schedule. Usually asynchronous LUN-level replication is configured for a window of 15 minutes or 30 minutes.

This replication frequency schedule should be dictated by the RPO. If the LUN-level replication features on your SAN are not able to meet the RPO, then it's time to turn to SQL Server for assistance.

Care must be taken with LUN-level replication when used with SQL Server. If the various files that make up a SQL Server database are on different LUNs, then all of these LUNs should be replicated in the same consistency group. If not, the SQL Server database will probably fail to come online in the event of a failover because the different pieces are out of sync.

BC and Disaster Recovery by SQL Server

Out of the box, SQL Server has a number of great business continuity features that can be configured in an infinite number of ways to meet your needs.

Database and Transaction Log Backups

Depending on the RPO and RTO dictated by the business, extending SQL Server to meet the BC needs could be as simple as periodic backups of databases that are copied to a disaster recovery (DR) site.

Log Shipping

If the RPO window is very small, database transaction log shipping can be utilized. Every change that occurs in the database is logged to a transaction log, and this file can be automatically backed up, the file copied, and applied to a standby database on a second database instance at another site.

Log shipping is extremely robust and has been used by DBAs for decades to provide a simple and straightforward BC strategy.

Mirroring

If the log shipping does not provide a fast enough RPO, database mirroring can be leveraged to send a synchronous or an asynchronous stream of database changes to a second instance.

This mirror is configured per database, and can be quite powerful when implemented properly.

Note that database mirroring has been deprecated as of SQL Server 2012, and at some point will be removed from newer versions of SQL Server.

Availability Groups

SQL Server's flagship availability feature, "AlwaysOn" Availability Groups, can replicate data for both high availability (HA) and DR purposes, and can do so at up to eight additional copies of the data.

As with mirroring, synchronous or asynchronous data replications can be established to keep the RPO down to as low as seconds, based on the rate of data change and the bandwidth between the two SQL Servers. (However, this feature is available only in SQL Server Enterprise edition.)

Competing or Complementary?

Knowing the features available with both the storage layer and with SQL Server is only the beginning of the architectural decisions an organization must face when designing a solid business continuity plan for the data.

Both replication strategies present compelling arguments and solid features for use by a business to protect their data. Let your RPOs, RTOs, and in-house skill sets determine the appropriate technology for each application stack.

These features can be used in quite complementary manners, such as SAN LUN-level replication to a secondary array located in the public cloud. The VMs can be replicated between sites. Database data can be moved between servers, datacenters, or entire clouds completely transparent to the application. Even the file-system data from within the VMs can be replicated independently from other solutions. Use whichever combination of features best suit your organizational needs for business continuity for all of your needs.

Be careful about the technologies in use. For example, if LUN-level replication is in use, and a DBA enables database log shipping to reduce the failover time, the changed database data is being moved to the secondary site twice. This configuration consumes twice the WAN bandwidth, and can clog the connection, which will result in both data streams slowing down.

One consideration when architecting BC for data platforms is to determine which group (or groups) is responsible for the DR process. In some organizations, the infrastructure team is solely responsible for the DR architecture, and the SQL Server environments are treated like every other server. Data replication, failover and failback processes are identical with the other servers to keep the process as streamlined and as simple as possible.

Other environments let the different applications select the BC strategy that best suits the RPOs, RTOs, and skill sets of the administrators. If the RPO for the application-layer is good enough for the application itself but not the underlying data, a combination of strategies can meet the RPO and RTO for the entire application stack while keeping the architecture as simple as possible.

Up Next

The options and strategies for protecting the database are endless, so choose wisely. Next, we'll discuss SQL Server implementation features and how they relate to the storage architecture.

5

SQL Server Considerations

Successfully managing the intersection of SQL Server and storage is crucial for any organization. Knowledge of how SQL Server uses I/O for performance can help a storage administrator design the infrastructure so it improves the efficiency of the entire stack.

In this chapter, we'll explore number of tricks with storage can be leveraged to improve the performance or operational overhead of SQL Server.

SQL Server Licensing Reduction

The SQL Server licensing discussion is at the forefront of a CFO's mind during the yearly renewal cycle. While not as expensive as some other enterprise applications, SQL Server is one of the more expensive applications in an organization. The cost to license SQL Server could outweigh the cost of the physical server that it runs on, or even the cost of the SAN where its data is stored. Reducing the associated costs with SQL Server is of key interest to management.

Your Mileage May Vary

Every organization's relationship with Microsoft licensing is different. While the advice you find here can be used to save your organization money, please verify with your Microsoft licensing representative that your licensing arrangement lets you leverage these suggestions.

Two key strategies exist to save your organization money on licensing: virtualization and flash storage.

Assuming your SQL Servers are virtualized, let's talk tech.

Virtualizing Your Databases

If your organization does not have your SQL Servers virtualized by now, weigh the options and consider virtualizing the databases as soon as you can. The agility and flexibility virtualization brings can save your administrators countless hours over the course of a year by easing the pain of tasks such as backups, disaster recovery (DR), server maintenance, and capacity management.

SQL Server licensing is based on three models.

- **Server plus client access license (CAL).** Each operating system that runs SQL Server is licensed, and then a CAL is purchased for every named user or device that connects to the SQL Server. Depending on the scale of your environment, this model might save you money, but it is rarely used for large scale deployments. Both the Standard and Enterprise editions can be licensed in this manner.

- **Per virtual machine (by core).** Each virtual machine (VM) must be licensed by the number of cores assigned (minimum of four cores). Standard and Enterprise edition can each be licensed in this manner.

- **Per host (by core).** Each physical CPU core must be licensed on each virtualization host where the VM could migrate to and run from. Only Enterprise edition can be licensed in this manner. If the hosts are properly licensed along with Software Assurance, your organization is free to deploy as many SQL Server VMs as can fit onto the licensed hosts.

So, for example, let's assume that you have a decent number of SQL Server virtual machines.

The modern virtualization host is quite powerful. It contains a large number of physical CPU cores and memory. However, SQL Server, either physical or virtual, needs a large amount of memory to operate efficiently. This memory is used for the *buffer pool*, an in-memory read cache for commonly access data. Historically, this buffer pool was kept large as a means to improve the performance when the storage layer performed poorly.

When the administrator starts to pile SQL Server VMs onto a single host, usually the first thing that fills up is the host memory. The host CPUs are usually not under pressure at all, while the host is at the maximum number of VMs it can contain, all because the memory consumption is maxed.

Enter an all-flash SAN.

The storage is now extremely fast, and the latencies are very low. In some cases, SQL Server might not need to keep as much of this commonly-accessed data in memory because it can fetch it from the storage layer *fast enough* to maintain performance.

In this example, we are now in a position to reduce the memory assigned to the VM. If we can reduce the memory and maintain performance, the number of VMs on that host can go up. If this number can go up enough, we might be able to squeeze all of the SQL Server VMs comfortably on one less physical host.

Since we have the unlimited rights to deploy as many SQL Server VMs on the hosts we have licensed, we can now work to reduce the number of hosts that we need to license. If we can reduce just one host from the license footprint, the amount of savings for the organization could be as high as $125,000.

That's serious money.

SQL Server vs. Array Features

Sometimes SQL Server features can be quite complementary to the SAN's set of features. Other times, SQL Server features can wreck a SAN's performance.

Compression

SQL Server DBAs can leverage table-level row or page compression to reduce the data footprint stored on disk. This compression is even maintained when the data is loaded into memory for processing, increasing the amount of data that the SQL Server buffer pool memory can hold. The SQL Server CPUs are marginally busier, but the reduced I/O boosts the performance of the SQL Server database.

However, if compression and/or deduplication is enabled underneath the SQL Server data, compressed data is not really further compressible. All it does is create a much greater workload for the data savings features on the SAN. The net results are greater SAN controller CPU consumption, higher latencies for the SQL Server storage, and much lower data savings rates on the physical storage.

> **Layers and Roles**
>
> Be sure to discuss with all parties involved where compression and deduplication is more effective — at the SQL Server or at the SAN. Usually, the SAN is more efficient at managing the deduplication, and can achieve greater data savings.

Transparent Data Encryption

SQL Server's Transparent Data Encryption (TDE) allows administrators to configure a very robust database-level encryption that encrypts the database on disk. Security is absolutely vital to the business, and encrypting the data at the database level is required in some organizations.

As with compression inside the SQL Server database, TDE can also cause the data savings percentages on disk to plummet and the workload on the SAN to rise.

Most SANs contain a feature to enable encryption of the storage layer, thus enabling data encryption at rest. If the organization security policy allows, consider encrypting the data on disk rather than inside the SQL Server engine. The security at the SAN layer could be enough to satisfy organizational requirements, and the data savings will allow much more data to be placed on the SAN efficiently.

> **Encryption Layers**
>
> Keep in mind that this LUN-level encryption is at the SAN level only. If the VM is backed up or cloned, and this secondary copy is placed on unencrypted storage, the security layer is removed. Encryption at the database layer moves with the VM through the data management life cycle.

Partitioning and File Groups

Some SQL Server tables can be quite large. Billions of records can be stored in a table. Queries can slam these tables hard, causing full table scans to funnel down one table and into one path to the storage.

SQL Server can leverage table-level partitioning to reduce the amount of data that a query will have to scan to find the necessary records. Data can be grouped by whatever logical grouping that best fits the data. A partition could be defined on a monthly basis for order history, and a partition would exist for each month that an order has been recorded. (**Figure 5-1**)

The query engine can then align the I/O requests across only the partitions which contain the data that the query requests.

Commonly joined tables are normally placed in the same database filegroup and data file, which forces the storage engine to funnel the requests down one path to disk.

Why does this matter with regards to storage?

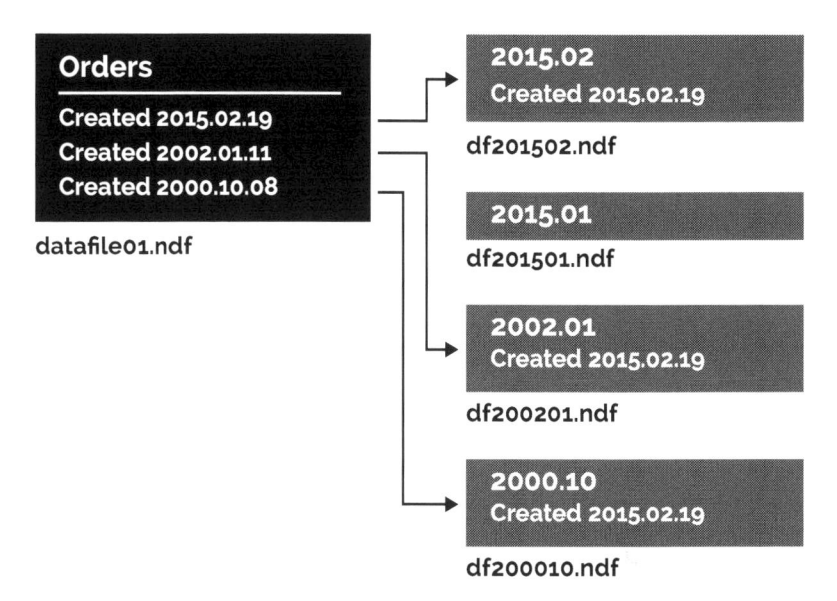

```
┌─────────────────────────┐         ┌─────────────────────────┐
│ Orders                  │    ┌───►│ 2015.02                 │
│ ─────────────────────── │    │    │ Created 2015.02.19      │
│ Created 2015.02.19      │────┘    └─────────────────────────┘
│ Created 2002.01.11      │──┐       df201502.ndf
│ Created 2000.10.08      │  │      ┌─────────────────────────┐
│                         │  │      │ 2015.01                 │
│ datafile01.ndf          │  │      └─────────────────────────┘
└─────────────────────────┘  │       df201501.ndf
                             │      ┌─────────────────────────┐
                             └────► │ 2002.01                 │
                                    │ Created 2015.02.19      │
                                    └─────────────────────────┘
                                     df200201.ndf
                                    ┌─────────────────────────┐
                                    │ 2000.10                 │
                                    │ Created 2015.02.19      │
                                    └─────────────────────────┘
                                     df200010.ndf
```

Figure 5-1: Spreading out the database workload

The files underneath database partitions or filegroups can be placed on multiple drives, corresponding to multiple disks, SAN LUNs, active controllers, and disk pools.

SQL Server will better parallelize the I/O operations within the SQL Server engine. The overall performance of the I/O task can be increased by spreading out the corresponding workload underneath SQL Server, reducing the impact of infrastructure or SAN controller or disk pool bottlenecks. This configuration can make the entire I/O operation more efficient (**Figure 5-2**).

Buffer Pool Extensions

SQL Server can leverage a buffer pool extensions (BPE) file placed on an SSD-housed drive as a spill location for memory pressure.

Multiple I/O Streams

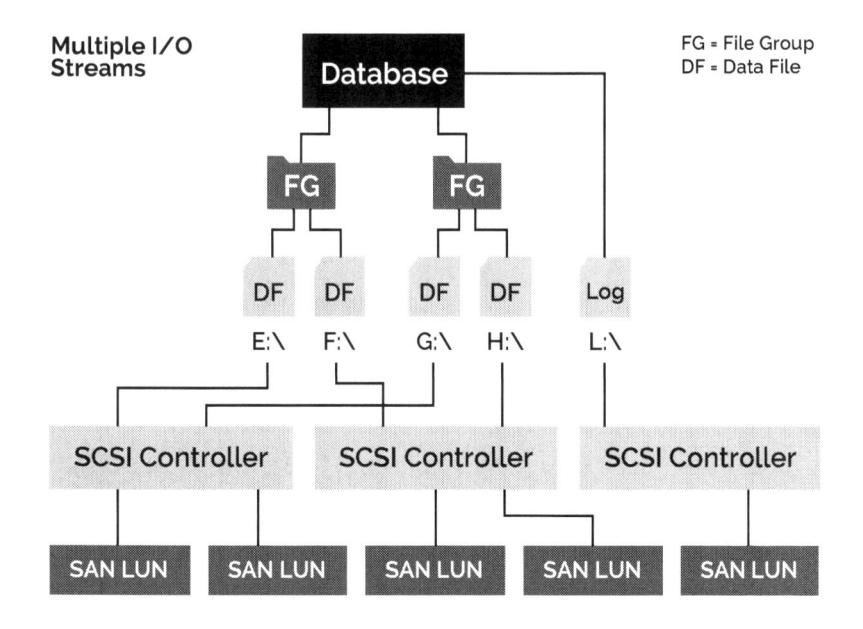

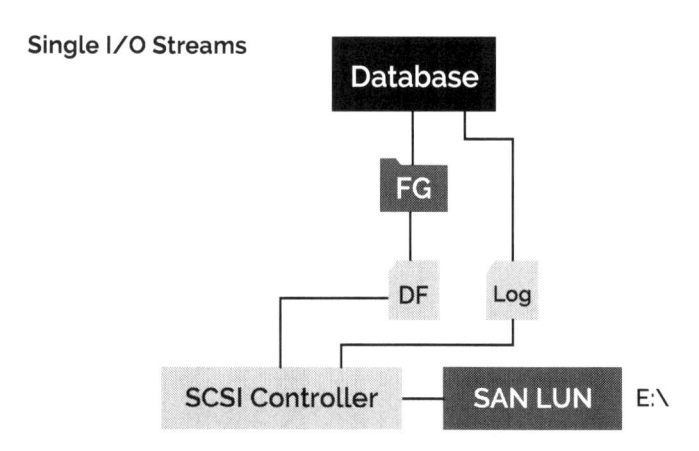

FG = File Group
DF = Data File

[Diagram: Database → FG, FG, Log; FGs → DF (E:\), DF (F:\), DF (G:\), DF (H:\); Log → L:\; connected to three SCSI Controllers and five SAN LUNs]

Single I/O Streams

[Diagram: Database → FG, Log; FG → DF; DF and Log → SCSI Controller → SAN LUN (E:\)]

Figure 5-2: SQL Server workload spread

If the SQL Server is under memory pressure and a BPE file has been defined, SQL Server can leverage this file and its underlying speed to extend the assigned memory to improve performance. This file can then be placed on an SSD-tier of a SAN, and improve the performance of the SQL Server.

Monitoring, Management, and Support

Monitoring

Monitoring storage performance must come from many levels. Monitoring utilities that collect and analyze data across the platforms is key to understanding the cause-and-effect relationship that one infrastructure component has with other components.

Education is also important. Personnel from one team might see performance metrics from another infrastructure layer and not understand how to interpret its true meaning, which can lead to a proactive warning being ignored and the situation turning into a reactive support call in the middle of the night because of a critical system failure.

Management

Managing SQL Server in an enterprise takes a conscious effort to involve representatives from all layers of the infrastructure and application stack, including:

- SAN

- Interconnect

- Virtualization (if applicable)

- Server / OS

- SQL Server

- Application owners

Working together to set performance SLAs, availability requirements, DR RPO and RTOs, and support statements is critical to properly managing these platforms. Without effective communication between the teams, any performance or availability anomaly will result in endless finger-pointing and the blame game, and this is counter-productive for the business.

When each team understands the high-level details about the other team's responsibilities and platforms, the technical challenges are guaranteed to be resolved quickly and easily.

Support

Validating your SQL Server configuration with your storage vendor's official SQL Server reference architecture is also recommended. The SAN vendor might have special settings or best practices that you should follow from the start that can help improve the performance and availability of the SQL Server on their storage platform.

Up Next

SQL Server is a lot more complex than many administrators realize, but with a solid understanding of its features and capabilities, architects can improve the power of the entire stack and the speed at which the business can operate, all while reducing the operational costs of the platform. Next we'll discuss specific recommendations for the SQL Server and storage layers that can squeeze the most performance from your environment.

6

Storage Best Practices

Now that we have shown you how important storage performance is to SQL Server performance, how can you make the most of your storage performance?

We first need to see if something else in the infrastructure is artificially slowing down the SAN's performance. Once we rectify any infrastructure bottlenecks on the way to the storage, then it is time to validate the SAN configuration itself.

Certain settings and configurations can be tweaked to improve performance at each layer of the infrastructure, and, depending on the type of SAN in place, these details can make your SQL Servers all-star performers.

Best Practices Start Early

Fixing storage misconfigurations and pain points start as soon as you get the array in the rack. Some of the misconfigurations and challenges are infinitely easier to detect and fix before you start putting your data onto the storage for production purposes. Thoroughly test any new storage purchase to ensure you are receiving the expected performance from the device before it goes into production.

Hardware and Bottleneck Detection

Imagine that you just received a brand new Italian exotic sports car as a birthday present from a rich relative that you never knew existed. It looks beautiful and sounds great when you start it up, but when you try to drive it, it will not go above walking speeds. You take it to a specialist mechanic who spends thousands of your dollars looking at the engine, but cannot seem to find anything wrong. Then you walk around the car to find that you have a flat tire. At least that is easy to fix, right?

The same situation happens all the time with bottlenecks inside the infrastructure. One large pain point between the data and the application can cause the entire system to slow to a crawl. The viewpoint from one layer of the system might not reveal the root cause, because the symptoms of the problem can change from each perspective. The underlying problem could be the same. Misconfigurations can, and do, cause these sort of problems in the infrastructure.

Many go largely undiagnosed because the teams who manage the layers do not recognize that the symptoms other teams report may be initiated within their own layer.

Take the high-level holistic view of the environment and start to drill into each layer and put performance metrics at each point along the way.

Storage Bottlenecks

Figure 6-1 shows some of the potential storage bottleneck points along the path from disk to the SQL Server.

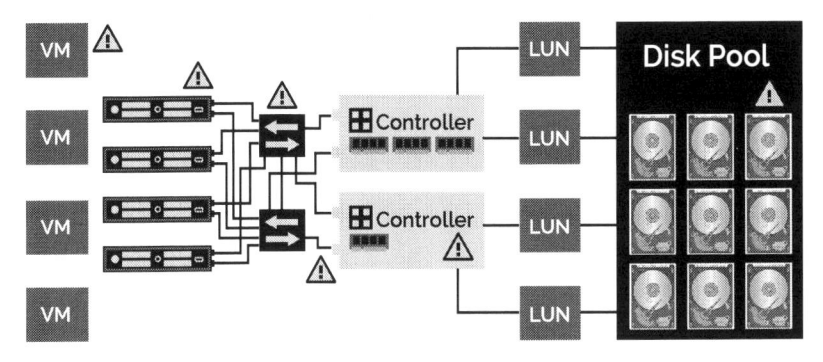

Figure 6-1: Storage bottlenecks

These bottlenecks could be due to a number of things.

- The disk pool or RAID configuration in the SAN could simply be slower than what the SQL Server workload demands.

- The SAN controller could have too small of a cache to adequately buffer the workload.

- The controller's port to the storage interconnect could also be misconfigured or at its peak capacity.

- The interconnect switch or switches could be misconfigured or at peak capacity, or even slower than the rest of the interconnects — which effectively slows the entire path down.

- The interconnect port on the physical server could be at peak capacity or have a misconfiguration.

Additionally, if a hypervisor is being used, several layers could be misconfigured or mismanaged, and can slow down the I/O request even further.

Operating System Bottlenecks

Once you get inside the operating system, even more bottlenecks can exist. They can even exist at the SCSI controller level (physical or virtual) within Windows Server, inside the SQL Server engine, or inside the database data files (**Figure 6-2**).

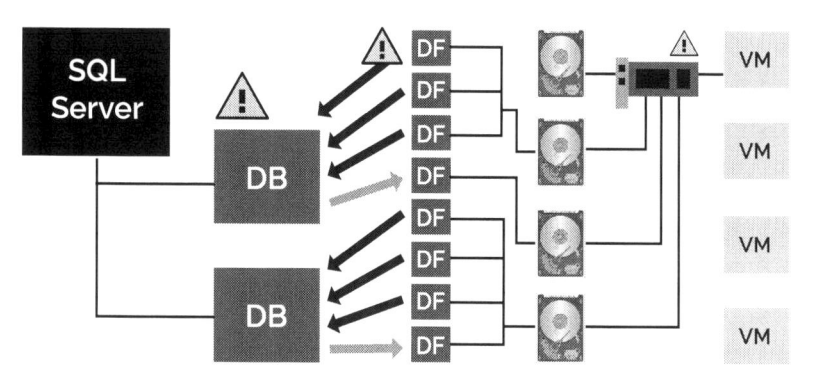

Figure 6-2: Operating System bottlenecks

Ordinarily, the end result of this series of challenges is the latency being slower within SQL Server and Windows Server than at the SAN. So, the storage administrators see very good performance and low latencies, but the DBAs see very poor performance and very high latencies.

The strategy to resolve this challenge is with performance data. At each and every stop between the two sides, measure the appropriate performance metrics and start to put the same type of metric in place end-to-end. If the numbers differ wildly between two points, stop and drill into that portion of the topology to determine where there's room for improvement at that point in the infrastructure.

For example, if the SAN reports 1.5ms response time for read operations on a given LUN, and the hypervisor shows about the same value to the same LUN at the same point in time, then the interconnect layer is probably in good order.

However, if the Windows layer inside the VM that sits on the LUN reports a read latency of 25ms at the same point in time, check the hypervisor and Windows-level disk configurations.

Objective numbers do not lie, and they can help resolve disagreements between teams. The goal is to have the best possible performance of the application (and all layers beneath it) to provide the business with the best possible tools to excel in the face of stiff competition.

Work together, put the numbers to each layer, and let the performance values drive the investigation.

Block Size

Block size is a very overlooked portion of tuning the environment to improve the storage performance. Block size is the group of contiguous space used to manage data placement on disk. Each layer of storage, from Windows Server, to the SAN LUN architecture, to the SAN disk configuration, manages data placement in these blocks.

For example, Windows Server's default block size is 4KB. Using round numbers for the example, if the C: drive is 100GB in size, it is broken into over 26 million logical units of 4KB per unit.

Windows Server 2008 and later also use a 1MB partition offset so that the partition starts far enough into the disk to avoid any strange block sizes or offsets from the storage layer.

Microsoft's Recommendations

Microsoft tells SQL Server DBAs to ensure that all Windows Server logical drives are formatted for a 1MB partition starting offset and a 64KB NTFS allocation unit block size.

New GPT-formatted partitions have a 128MB starting offset.

As long as the starting offset is a multiple of 64KB, you are probably aligned with your storage architecture. SQL Server workloads generally read and write data in multiples of 64KB, so if the partition is aligned properly, SQL Server and Windows will align and maintain the disk performance.

The SAN disk and LUN configuration can also be aligned on a block size. What if these block sizes are different within Windows from what the SAN LUN is configured for?

Windows Server

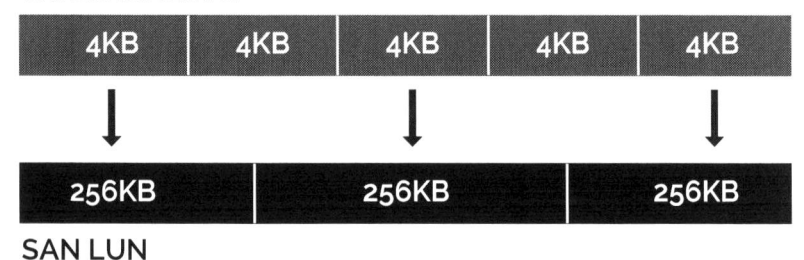

| 4KB | 4KB | 4KB | 4KB | 4KB |

| 256KB | 256KB | 256KB |

SAN LUN

Figure 6-3: Block size mismatch

In **Figure 6-3**, Windows Server NTFS is configured for a 4KB block size. The SAN LUN is configured for a 256KB block size. If the workload on the Windows layer is random, a single 4KB block read could force the SAN to read 256KB of data just to fetch the

4KB and return it back to Windows. Now repeat this millions of times over the course of a day for a normal server. The storage layer might be reading much more data from disk than what is necessary, and the result is slower overall performance for the workload.

As a result, consult with your storage vendor's validated SQL Server reference architecture for the recommended LUN block size. Configure the SAN LUN per their recommendations, and then test to verify performance.

For most workloads and environments, a 64KB block alignment strategy is ideal within the SQL Server and Windows layers. Occasionally, an alternate block size might provide slight performance improvements to SQL Server, but extensive testing should be performed to come to that determination.

Tuning for Flash

All-flash SANs provide extremely low latencies when compared to their spindle counterparts. This low latency will most likely help improve the SQL Server performance if the storage layer is a current bottleneck to performance (most of the time, it is).

However, simply moving your SQL Server databases to an all-flash SAN might not make as large of a performance improvement as you might think.

To determine this, start by reviewing the infrastructure, looking for additional bottlenecks that are preventing the all-flash SAN from kicking into high gear and flexing its muscles.

Once the infrastructure bottlenecks are removed, it's time to dive into SQL Server storage internals.

Spreading out the workload could make an improvement to overall performance. If the storage layer is no longer the performance bottleneck, the bottleneck shifts farther up the infrastructure stack. It could even make it all the way into the Windows OS layer. You'll see a symptom of this challenge when the SAN performance latency metrics are quite low but the Windows layer Perfmon metrics for disk latency are quite high.

If the interconnect layer has been ruled out as a problem, the Windows I/O queues per SCSI controller could actually become the bottleneck. Adding additional disks with additional SCSI controllers, each with their respective queues, could relieve some of this pressure and reduce the latencies.

SQL Server can only make changes to data (insert, update, and delete operations) as fast as it can write a record of that change to the transaction log file. All-flash can make the latency of this operation very small, improving the performance of the change operation. This log stream is in a sequential write stream to disk.

Dedicating a SCSI controller and separate disk or LUN for database logs can help with the performance if the volume of writes is too high.

TempDB is the SQL Server container for all of the random tasks that occur on a routine basis, such as sorting, temporary objects, spooling, and database integrity checks. Placing this database's data and log files on the all-flash array can also improve the performance of any operation, read and write, that makes heavy use of TempDB.

Contrary to common belief, database index maintenance is still critical to keeping a healthy database. Proper maintenance keeps the database layer efficient. Moving the database indexes to a different file group and data file, and placing this file on high performance media and LUN could also improve the performance.

SQL Server Storage Requirements and Budget

Most data volumes grow exponentially each year. Your organization is sure to be no different; businesses collect more data from more sources and want this data online for longer to help gain fresh business insight. When businesses plan their data and growth strategies, the common process is to project the amount of space needed a certain number of years in the future.

Capacity is only half of the challenge. The storage performance is also key to maintaining that data. Trending the performance metrics next to the data growth rates will help you determine if your storage can deliver the necessary performance down the road.

For example, **Figure 6-4** shows a sample projection from a business. This business has been rapidly adding data into their environment over the last 5 years. Recently, they purchased a new SAN that they expect to last for the next hardware cycle of 5 years. It contains 80 TB of usable space. In their projections, they will easily be able to store their data within this footprint through 2020, when the next purchase cycle is due.

However, the SAN they purchased can sustain a maximum I/O per second (IOPs) of 500,000.

Storage Consumption and Performance Estimates

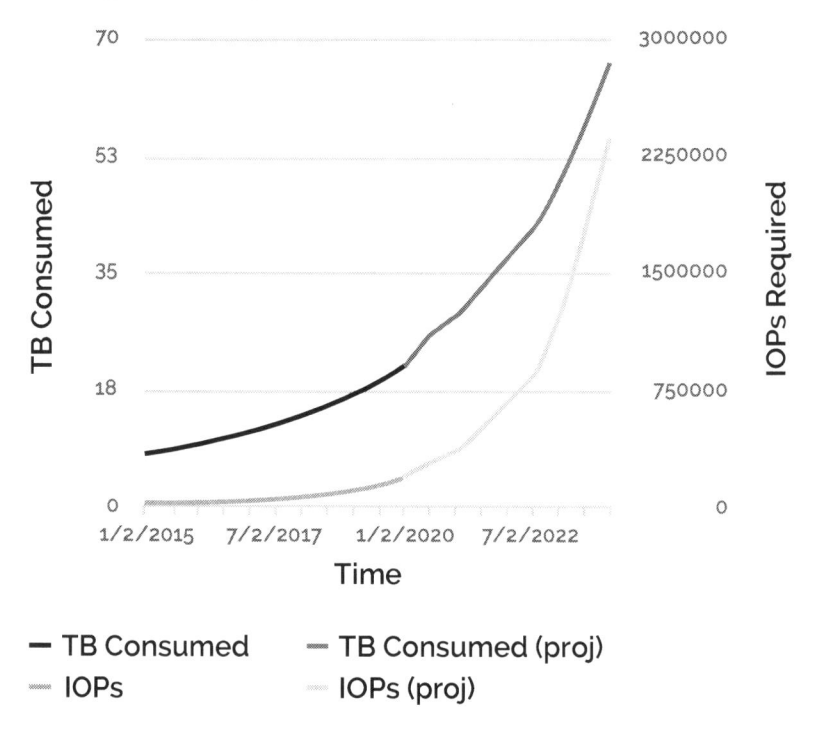

Figure 6-4: Example capacity projection

Based on the historical data and projections, the storage will effectively run out of performance around the middle of 2017. Beyond that point, the performance of the array will suffer as data continues to be added, and the business could be negatively impacted by this upcoming performance bottleneck.

You should measure the ongoing performance metrics from within the different infrastructure layers around the clock, and use these to project the storage performance requirements (not just the capacity requirements). When budgeting for the next storage purchase, take into consideration the performance necessary to maintain the business-critical processes that run on the storage. Purchase the fastest array that the business can afford that can meet these requirements.

Up Next

Now that you know how to design and tune the storage layer for maximum performance, let's focus on keeping you current. Newer SQL Server versions can be used to improve the efficiency and performance of the storage layer. In the next chapter, you will learn about some features of SQL Server that can be used in conjunction with your storage to create more benefit for the business.

Modernizing SQL Server

Today is the right time to consider upgrading your SQL Servers to the most recent version, as well as improving the infrastructure underneath it to provide the foundation for peak performance from the applications.

Features in the Latest Version

In-Memory OLTP

SQL Server 2014 released In-Memory OLTP, a new feature that allows the database administrator (DBA) to move individual tables to a new in-memory layer. The performance boost can be quite substantial (as much as 30 times faster), but this feature is quite dependent on the storage layer.

While the working set of data stays resident in memory, any changes must be written to disk before the operation completes. The faster the storage layer, the faster the operation completes, and the faster the application continues with its work.

Furthermore, if a SQL Server needs to restart, all of the data in these in-memory tables must be read from disk before the database is ready to use. If hundreds of gigabytes (or more) of data must be read, this operation could take a lot of time on slower storage, leaving the business with an extended outage in the meantime.

As a result, if your organization is considering adopting In-Memory OLTP for your applications, ensure that you have the fastest possible storage layer that your business is willing to afford.

Stretch Database

Starting with SQL Server 2016, SQL Server can store portions of a database in Azure. The beauty of this new feature is that an organization can select tables that contain legacy or archival data to be pushed into Azure. An organization can now leverage the fastest tier of storage in their internal datacenter for the data that is accessed the most, while leveraging the cost-effectiveness of Azure for storing the data that is accessed the least.

Azure Integration

SQL Server 2016 will contain a number of features designed to leverage the power of Azure, in addition to the Stretch Database feature.

Azure SQL Database can now support In-Memory OLTP and real-time Operational Analytics. Databases can now be directly backed up or restored from Azure. Availability Group secondary replicas can now be placed directly within Azure. Transactional Replication directly to Azure SQL Database is supported as well.

SQL Server 2016 and Azure SQL Database continue to grow in their power, and provide even greater numbers of options for your organization to power your business.

Buffer Pool Extensions

SQL Server 2014 also added Buffer Pool Extensions (BPEs) to let the administrator define a flash device to extend the memory allocation on the server. This feature allows SQL Server to use less memory while leveraging flash storage to maintain or improve performance. This feature is quite useful in a virtualization consolidation exercise where the memory footprint of the virtual machines (VMs) needs to be reduced.

Clustered Columnstore Indexes

Designed for OLTP tables with more than 10 million rows of data, columnstore indexes create indexes column-by-column instead of row-by-row, which allows developers to produce data warehouse-style analytical queries on OLTP workloads.

SQL Server 2016 improves upon the columnstore index feature by adding secondary regular indexes in addition to the updatable columnstore index feature that was introduced in SQL Server 2014.

SQL Server 2016 also allows the modification of a non-clustered columnstore index, which improves the flexibility of the feature for developers looking to improve the performance of their analytics queries.

SMB3-Based Network Shares

SQL Server even allows for the placement of the databases on an SMB3-based network share to improve the availability of the databases. This feature is useful for central database storage with SQL Server Failover Cluster Instances (FCIs).

If your SAN supports the SMB3 storage protocol, consider using SMB3 as a target for your shared storage instead of traditional shared LUNs or iSCSI targets. This can reduce the complexity of your SQL Server high-availability solution.

Benefit Analysis of Upgrading

Support

First and foremost, if you are currently running older versions of SQL Server, check to see if your version is covered by Extended Support.

- **SQL Server 2008 and 2008R2.** Mainstream support for SQL Server 2008 and 2008R2 ended in 2014, and Extended Support will end in 2019. That will be here before you know it.

- **SQL Server 2005.** If you have a 2005 instance in your environment, Extended Support ends in April of 2016.

- **SQL Server 2000 and older.** If you are unfortunate enough to have a 2000 or older version of SQL Server in your environment, the extended support date passed in 2013, and you should drop what you're doing and work to get these instances upgraded as soon as possible.

Expired support means that if you have any production outage, encounter bugs, or have general help, Microsoft could refuse to help you. This is not a position that your organization should ever be in. The upgrade path is straightforward. You will need to work with your business to demonstrate the value in the upgrade.

New Features

If your business needs additional reasons to upgrade legacy installations, just look at the new and updated features that come with the latest version of SQL Server.

- "AlwaysOn" Availability Groups for improved high availability and disaster recovery
- Stretch Database for offloading archival data to Azure SQL Database
- In-Memory OLTP for greatly improved performance
- Improved Columnstore index features
- Improved query processing and optimization
- Backup encryption support
- Support for greater numbers of CPU and memory
- Improved security
- Improved T-SQL functionality

Widespread Business Benefits

The added power provided by the newest version of SQL Server is sure to improve the performance and availability of your database servers. This means that productivity will increase. Data availability will improve. Administration overhead will decrease. Security will improve. The business benefits of the upgrade will surely be felt by all.

Plus, if your environment has the numbers of SQL Server instances that tip the scales for license reductions through virtualization, you can receive all of the features listed above while reducing the ongoing operational costs. Save the business money while improving the performance and availability!

Maximizing SQL Server with Flash

All-flash SANs present a two-fold challenge. The upfront costs of the SAN can be daunting for some organizations, and the business case must be made in order to justify the purchase. Once acquired, the environment must be adjusted to make the most of this new storage layer.

Perception and Justification

One substantial challenge that you need to overcome within your organization is the perceived separation of dependencies between SQL Server and storage.

To help show the benefit of all-flash storage to the DBAs and application owners, you need to turn an existing performance challenge with your existing storage into a business proposition.

Management generally goes cross-eyed if you start to describe challenges such as "thin provisioned storage," "fragmentation," "block sizes," or "fast" or "slow" without quantifying it all. Some might feel that you just want a shiny new toy to tinker with.

You must *put the business first* in the proposition.

Examples of extremely compelling reasons may include:

- "The business users are not satisfied with the speed of the application, and the storage is holding back the application."

- "Each user spends up to two full hours each day waiting on the system to execute a task, and 80% of these delays are because the system is waiting on storage."

- "Faster storage will improve user productivity by up to 1.5 hours for each employee, each day."

The justification you present should show the value of the storage to the business in terms of lost productivity. A good business proposition can help persuade the organization to invest in appropriately fast storage.

Adoption

Once acquired, a flash storage array can be installed and your data can be moved over. At this point, the infrastructure bottlenecks will shift. How they shift depends on your topology and environment, but they are guaranteed to shift. Network adapters can now reach peak throughput, causing a slowdown. HBAs can instigate latency at the server through queue bottlenecks, which the DBA will detect and have contradictory performance results than what the SAN reports.

At this point, you should be ready to review the tips presented in this book and be able to work to correct any internal bottlenecks. Your goal is to shift the bottlenecks back onto the SQL Server database and application layers, and not have the infrastructure hold either of these back.

That's a Wrap!

You have now been introduced to SQL Server and how it uses storage, its workload properties, and how to tune the storage to make the most performance for the database.

We hope your journey here has helped you to better understand the intersection of these two technologies and how one layer matters to the other.

You can continue to learn much more about SQL Server and storage performance and availability topics by regularly visiting *www.tegile.com/sql* and *www.gorilla.guide*.